Backpack Bear's

Invertebrates Book

Includes Spiders and Insects

Written by Alice O. Shepard

Starfall®

www.**Starfall**.com

ISBN: 978-1-59577-089-9

Starfall Education, P.O. Box 359, Boulder, CO 80306

Table of Contents

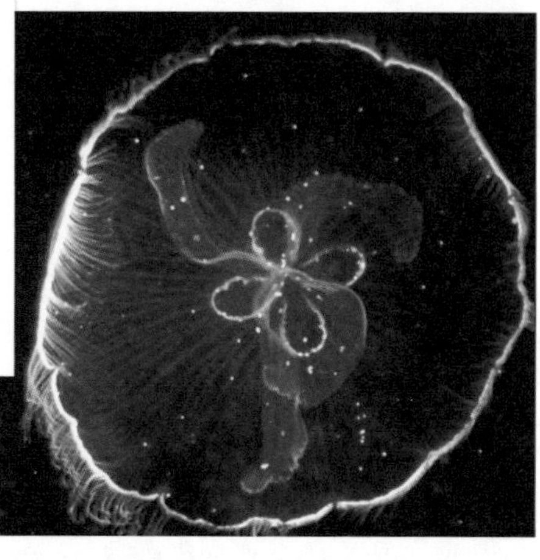
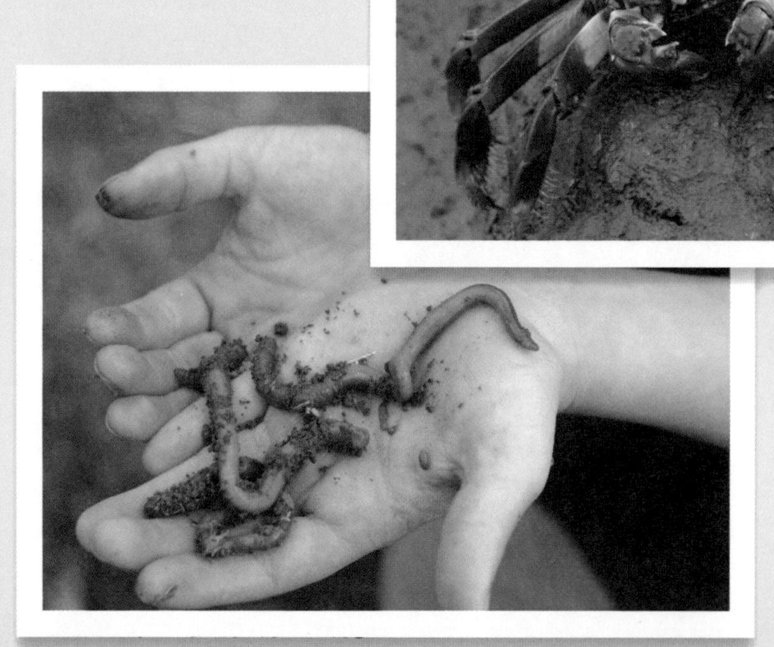

Invertebrates

These animals may look different from each other, but they all have one thing in common. They are ***invertebrates***!

Invertebrates are animals without backbones.

Did you know that all invertebrates are "cold-blooded?"

You can ***observe*** invertebrates almost anywhere! You will see them crawling, wiggling, jumping, swimming, flying, and oozing. Some barely move at all!

Have you seen these animals before? They are all invertebrates.

By observing invertebrates carefully, you can learn how to sort them into groups.

Invertebrates (Animals withou

These are different groups of invertebrates. Arthropods are the largest group.

Jointed Legs
(Arthropods)

Backbones)

Soft Bodies
(Mollusks)

Spiny Skins

Stinging Tentacles

Natural Sponges

Worms

Crab

Arthropods: Invertebrates with Jointed Legs

Crabs, spiders, and honeybees are invertebrates, but they also share another characteristic. Can you see what it is? They all have jointed legs! Scientists named this group ***arthropods***. Arthropod means "jointed foot."

Arthropods have a hard outer skin called an ***exoskeleton***. Their exoskeletons protect their bodies and give them their shape.

Spider

Honeybee

Crab

Cicad

Arthropods do not grow like you do. As *you* grow, your skin stretches to make room for your bigger body, but an arthropod's skin is hard—it can not stretch!

Arthropods shed their exoskeletons as they grow new, bigger ones.

Stink Bug

This ant and spider are both arthropods. How can we tell? They both have jointed legs!

Look closely. These arthropods are not the same.

- Ants have six legs and three body parts.
- Spiders have eight legs and two body parts.

Ants belong to a group of arthropods called ***insects***. All adult insects have exactly six legs and three body parts. A spider is not an insect because it has eight legs and two body parts.

Ant

Spider

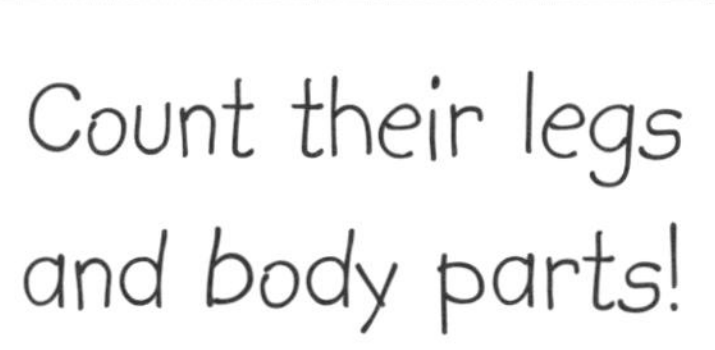

Insects:
The Largest Group of Arthropods

There are many, *many* different kinds of insects, more kinds than all the other animals in the world, combined!

Sugar Maple Borer

Small Tortoiseshell Butterfly

Feather-horn Beetle

Banded Alder Borer

Longhorn Beetle

You probably remember that adult insects have six legs and three body parts, but did you know that insects also have two ***antennae*** at the top of their heads?

Sometimes, antennae are called "feelers," but they can do more than feel. Insects use their antennae to smell and taste, too!

Dragonfly eyes

Can you guess how many eyes this insect has? No, it doesn't have just two. This insect has thousands of eyes!

Each of the big eyes you see in this picture is made up of thousands of tiny eyes.

With these eyes, this insect can see things that move really fast!

Insects have clever ways to protect themselves from predators. Some are poisonous. Others sting. Most hop, crawl, or fly away.

Many insects have excellent camouflage. Can you find the insects hiding in the pictures below?

Walking Stick Insect

Leaf Insect

Grasshopper

Butterfly

What do you think would happen if a predator tried to eat this insect?

Spiny Bush Cricket

Honeybee

A few kinds of insects, such as honeybees, live in a ***colony***. These insects work together to find food, build shelter, and take care of their babies.

Honeybee

Honeybees and some other insects do a very important job. They carry pollen from flower to flower. If they did not do this work, plants would not grow, and life on our planet would not survive!

Metamorphosis of a Monarch Butterfly

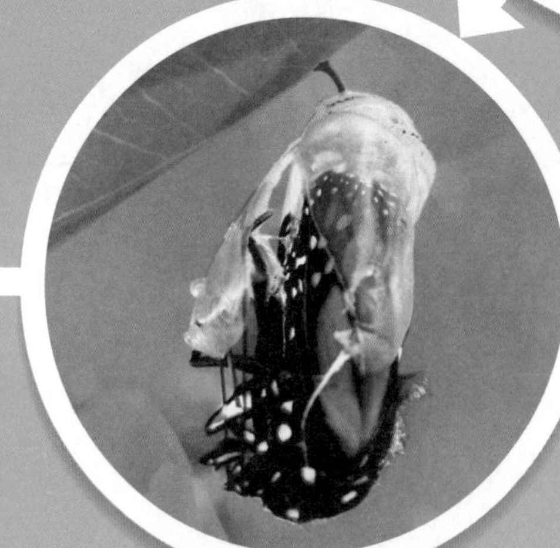

Almost all female insects lay eggs. Some hatch into worm-like babies called ***larvae***.

The larvae do not look like their parents. Their bodies change as they become adults. This transformation is called ***metamorphosis***.

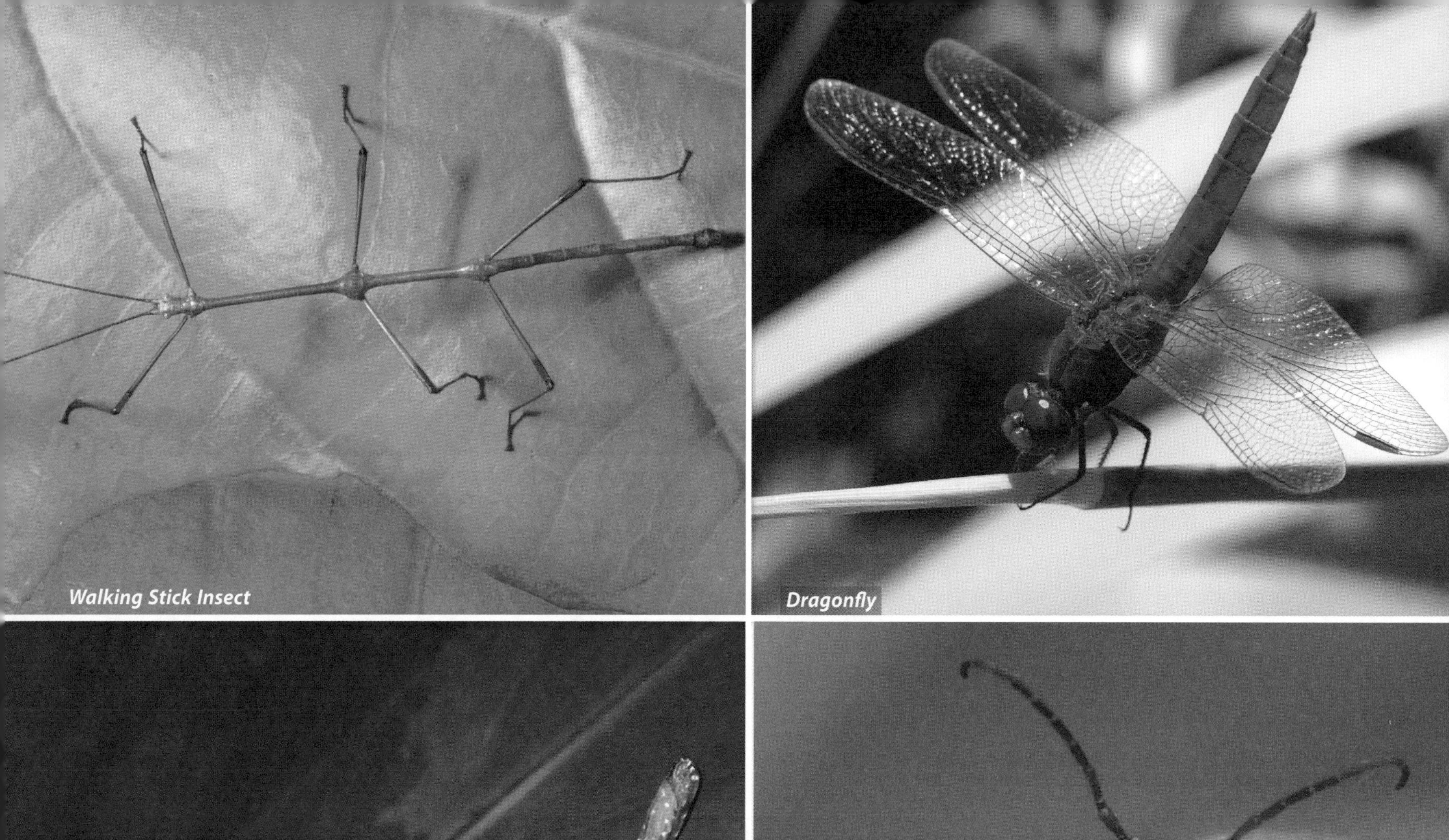

Walking Stick Insect

Dragonfly

Fulgorid

Milkweed Beetle

Insects are the largest group of animals on our planet. They can be found flying in the air, crawling above and under ground, and even swimming in the water.

They are unusual and beautiful animals.

Luna Moth

Glossary

Antennae: A pair of sensory organs of some arthropods, used to feel, hear, taste, and smell

Arthropod: An invertebrate with jointed legs and an exoskeleton

Colony: A group of insects that live and work together and depend on each other for survival

Exoskeleton: The hard protective outer shell of an arthropod

Insect: An arthropod with three body parts, six legs, and two antennae

Invertebrate: An animal without a backbone

Larvae: A worm-like stage in the life cycle of an insect

Metamorphosis: A major change in the form of some animals that happens as the animal becomes an adult

Observe: To pay close attention and notice details about what you see

What Is the Food Chain?

Every living thing needs energy to live. Plants get energy from the sun, some animals eat plants, and some animals eat other animals. Whenever something that was alive dies, it returns to the soil. The soil helps new plants grow. Each living thing depends on other living things. This is called the Food Chain.

If you take out a link in the chain, you affect everything in that chain. For example, if you pollute the land and water, the plant life dies. Many animals depend on plants for food, and without plants they can not survive. Without the smaller animals to feed on, the larger animals disappear from the world.

Index

About the Author

Alice O. Shephard's favorite invertebrate is an arthropod with ten legs. Can you guess what it is? It is a crab! She likes them so much because they are one of her favorite things to eat, but only on special occasions.

Acknowledgements

Special thanks to Ken Holsher, Department of Entomology, Iowa State University, for helping us check this book for accuracy.

Photo Credits

The following images licensed from iStockphoto: *Girl and Snail,* on page 4, © Anna Karwowska; *Boys and Worms*, on page 4, © Pattie Calfy; *Girl and Ladybug*, on page 4, © ados; *Holding Starfish*, on page 5, © Elizabeth Bruders; *Octopus Under Surface*, on page 6, © Herve Lavigne; *Slimy Snail*, on page 7, © Steven van Niederhausern; *Tarantula*, on page 7, © John Bell; *Monarch Butterfly*, on page 8, © Lowell Gordon; *Ant*, on page 15, © Tomasz Zachariasz; *Ladybird*, on page 16, © Olga Khoroshunova; *Beetle*, on page 16, © Chartchai Meesangnin; *Green Flower Beetle*, on page 16, © Steve Smith; *Blue Beetle*, on page 16, © pkruger; *Dragonfly*, on page 16, © Alexander Chelmodeev; *Bumblebee*, on page 17, © Natasha Litova; *Mantis*, on page 17, © Evgenty Ayupov; *Common Metallic Longhorn*, on page 19, © Ivan van Rensburg; *Spiny Bush Cricket*, on page 23, © Morley Read; *Monarch Chrysalis*, on page 26, © Cathy Keifer.

The following images are used by permission of these photographers from Flickr.com: *Shell*, on pages 4 and 9, © Emma Murray; *Snail*, on page 9, © Ken Slade; *Cicada*, on page 12, © Amber M. Lujan; *Crab*, on page 12, © Donna Bjordahl; *Stinkbug*, on page 13, © Wolfgang; *Velvet Ant*, on page 16, © M. Howard Stephens; *Swallowtail Butterfly*, on page 17, © Jim Hughes; *Feather-horn Beetle*, on page 18, © Susan Sneath; *Banded Alder Borer*, on page 18, © Larry Zimmer; *Dragonfly Eyes*, on page 20, © Henrich Stabell; *Grasshopper*, on page 23, © Geoff Gallice; *Monarch Caterpillar*, on page 26, © Sandy Austin; *Butterfly Emerging*, on page 26, © Pieceoflace Photography, *Monarch Caterpillar*, on page 26, © 2007 Antoine Hnain.

Luna Moth, on page 29, © *Andrey Antov*. Used with permission. *Jellyfish*, on page 7, and *Dragonfly*, on page 28, are public domain photos from Wikipedia Commons. *Moon Jellyfish* on page 4, is from NOAA's Coral Kingdom Collection, photographed by the Florida Keys National Marine Sanctuary Staff.